IF DINOSAURS WERE HERE TODAY
THE HUNTED

Copyright © 2024 Bright Bound Ltd

First published in 2024 by Hungry Tomato Ltd
F15, Old Bakery Studios, Blewetts Wharf, Malpas Road, Truro, Cornwall,
TR1 1QH, UK.

No part of this publication may be reproduced, stored in a retrieval system, or transmitted in any form or by any means, electronic, mechanical, photocopying, recording, or otherwise, without prior written permission of the copyright owner.

A CIP catalogue record for this book is available from the British Library.

ISBN 9781916598942

Printed in China

Discover more at
www.hungrytomato.com

Picture Credits:

(abbreviations: t=top, b=bottom, m=middle, l=left, r=right, bg=background)

Corbis 1bg, 9tr, 20-21bg.
FLPA: Yva Momatiuk/John Eastcott/Minden Pictures 2bg, 9ml, 18-19bg, 25b.
Getty Images: Image Bank 28tl.
Natural History Museum London: De Agostini 26br.
NHPA: John Shaw 22tl; Nick Garbutt 3m, 5b, 7bl, 8br, 9br, 12-13bg, 14-15bg, 26tr.
Shutterstock: 4bg, 7mr, 8br, 9bl, 10-11bg, 16-17bg, 30bl, Alberto Andrei Ruso 7tl; Budmir Jevtic 6bl; Catmando 28bl; Daniel Eskridge 22b, 29tl; David.costa.art 6tl; Denis--S 6mr; DM7 23tl, 27tl; Dotted Yeti 23br; Elenarts 29mr; Ericreations 25tl; Frame stock footage 30tr; Gledriius 27mr; Ilker Murat Gurer 26ml; Johan Swanepool 28mr; Marina Summer 26bl; Natalia Van D 31mr; Penny Hicks 31br; Tatiana_Kashko_Photo 31tt; Ton Bangkeauw 24mr; Warpaint 29bl.

Every effort has been made to trace the copyright holders, and we apologise in advance for any unintentional omissions. We would be pleased to insert the appropriate acknowledgements in any subsequent edition of this publication.

IF DINOSAURS WERE HERE TODAY!
THE HUNTED

by John Allan
Illustrated by Simon Mendez

WARNING! These extinct beasts are not alive today. But just imagine if they were...

CONTENTS

The Story of the Dinosaurs	6
Timeline	8
Clever Decoy - *Ankylosaurus*	10
Herd Mentality - *Psittacosaurus*	12
Whipping Tail - *Plateosaurus*	14
Territorial Threat - *Anurognathus*	16
Charge! - *Styracosaurus*	18
Battering Ram - *Stegoceras*	20
Prehistoric Protection	22
The Lives of Dinosaurs Today	24
Did You Know?	26
True or False?	28
Uncovering the Past	30
Index & Glossary	32

Words in **BOLD** can be found in the glossary.

THE STORY OF THE DINOSAURS

Planet Earth is around 4.5 billion years old. Rocks containing traces of living things shows us that there's been life on Earth for around 3.6 billion years. During Earth's long history, the planet and the creatures that roam it have changed drastically. We've all heard of the dinosaurs, but where did they come from, and where are they now?

WHEN DINOSAURS ROAMED

Dinosaurs were the most famous and fascinating animals to come from these prehistoric times. Dinosaurs were the biggest land-living creatures to have ever lived. Alongside these giants lived smaller, bird-like dinosaurs, flying reptiles and huge ocean-dwelling beasts.

Then, 65 million years ago, the dinosaurs were suddenly gone! Scientists believe that a huge asteroid hit Earth, wiping out most living things. The extinction of the dinosaurs allowed for the rise of new animals: 4 million years ago, humans appeared!

FOSSIL FINDS

Humans began observing rocks and, in the 1700s, discovered that the **fossils** they contained were the remains of ancient plants and animals. Fossil hunting became popular and the study of fossils – **palaeontology** – was born.

In 1842, scientist Sir Richard Owen invented the term 'Dinosauria' to describe the giant creatures that had once walked the Earth. Their remains fascinated both scientists and ordinary people – everyone wanted to know what these creatures had been like.

THEN AND NOW

For over two centuries, dinosaurs have amazed and fascinated us. We wonder how they'd compare to the animals of today.

Could a Plateosaurus defend itself against the ferocious tiger and any other **carnivores** that try their luck?

Would horned dinosaurs happily live among other horned animals like the rhino or would they be enemies?

Would some dinosaurs find safety in numbers and remain in herds to help defend themselves from hungry predators?

THE UNKNOWN

We may never know exactly what it would be like to live with dinosaurs. We can only imagine, taking what we've discovered from their fossilised remains, and comparing it to what we know about modern animals to picture what life would be like if dinosaurs were here today!

If you've got the courage, read on...
...be prepared for some truly bizarre and spine-tingling - though imaginary - encounters between human or animal and beast.

TIMELINE

TRIASSIC PERIOD
(252–201 MILLION YEARS AGO)

Dinosaurs appeared towards the end of the Triassic **period**. They tended to live by the seaside, along riverbanks and in desert **oases** where water was plentiful. Early dinosaurs included Plateosaurus and Coelophysis.

CRETACEOUS PERIOD
(145–66 MILLION YEARS AGO)

This was when some of the most famous dinosaurs lived, including T.rex, Triceratops and Spinosaurus. Who knows what other dinosaurs would have lived since then if they hadn't all been wiped out by the huge meteorite?

JURASSIC PERIOD
(201–145 MILLION YEARS AGO)

During the Jurassic period, Earth's climate became moister and milder – new plants and forests grew, meaning new food sources for plant-eating dinos. As a result, both plant- and meat-eating dinosaurs started to grow much bigger.

PLATEOSAURUS

Name meaning 'broad lizard'.

Large and heavy, this plant-eating dinosaur belonged to the **prosauropod** group. Despite their big bodies, they could walk on their hind legs as well as on all fours.

ANUROGNATHUS

Name meaning 'jaws but no tail'.

This unusual flying reptile lived during the Age of Dinosaurs. It was small and easy to tell apart from others due to its lack of tail.

252 MILLION YEARS AGO

201 MILLION YEARS AGO

STEGOCERAS

Name meaning 'roof-horn' after its unusual skull.

This two-footed dino had a thick bony dome on the top of its skull which could've been used to charge at and fight off predators.

STYRACOSAURUS

Name meaning 'spiked lizard'.

This fierce-looking short-frilled dino had a long nose horn and horns around the edge of its neck frill.

MASS EXTINCTION

For millions of years, dinosaurs ruled the Earth, until there was a **mass extinction**. There is evidence that a **meteorite** struck Earth around 65 million years ago, exploding rock fragments, causing **tsunamis** and forest fires, resulting in the death of the dinosaurs and all other reptiles of the time.

PSITTACOSAURUS

Name meaning 'parrot lizard'.

This small, two-footed dino was named after the shape of its head – being square-shaped and with a heavy beak, it resembled modern day parrots.

ANKYLOSAURUS

Name meaning 'fused lizard'.

This heavily-built dinosaur had hard spikes along its back and a bony club on the end of its tail, making it well-protected from predators.

145 MILLION YEARS AGO ▷▷▷▷▷▷▷ **65 MILLION YEARS AGO**

CLEVER DECOY
ANKYLOSAURUS

Three playful sifakas spy a stranger in their thicket. Intrigued and a little intimidated, they throw themselves at the newcomer's head, taunting it and leaping away. But their target's not really the intruder's head but its tail! The animal is a **juvenile** Ankylosaurus who finds the attack of the sifakas hardly even an annoyance. It lumbers on regardless.

Ankylosaurus was well-suited to defend itself. Its wide body was covered with bony plates and spikes, and it had a heavy club on the end of its tail. As well as a weapon that could be swung at predators, the club may have acted as a decoy. Its resemblance to a head may have drawn an attack away from more vulnerable parts of its body. Ankylosaurus must've had strong muscles to move its heavy body. Today we'd probably harness that power; Ankylosaurus would make a great beast of burden, like modern donkeys, horses and sled dogs.

ANKYLOSAURUS
PRONOUNCED
an-kee-low-saw-rus

LIVED
Late Cretaceous period
70-65 million years ago.

LENGTH
up to 8 metres (26ft)

DIET
Herbivore

KILLER TAIL

Ankylosaurus was the biggest of the bone-plated dinosaurs. Its head was a solid plate of bone. Its back was covered in thickly-spiked plates right down to the end of its tail, where there was a club made from heavy chunks of bone. The tail was stiff and straight, like the shaft of a medieval mace. The club at the end could be swung with force against the legs of an attacker or to assert dominance over other Ankylosaurs.

PREHISTORIC TANK

Ankylosaurus' thick body was almost impossible for a predator to bite through. In a similar way, a tank's tough outer shell protects it from gun and missile fire.

HERD MENTALITY
PSITTACOSAURUS

A patter of three-toed feet on the turf. A flock of Psittacosaurus bunch together as predators circle. In a herd, they take protection from one another. One of their group miscalculates and splits away from the flock. In its Cretaceous home this would make it vulnerable, but here the 'menace' is from trained sheepdogs, and soon it's guided back to its companions.

It may well be that small plant-eating dinosaurs such as Psittacosaurus would be easy to breed, and so become important farm animals in the modern world. However, the herding instinct would still exist in them – it's a valuable survival mechanism. Threatened by meat-eating dinosaurs, plant-eaters like Psittacosaurus would find safety in a group. With lots of animals crowded together, perhaps only one would be killed by the attackers, and the rest would be safe.

PSITTACOSAURUS

PRONOUNCED
sih-tack-oh-saw-rus

LIVED
Early Cretaceous period
125-100 million years ago

LENGTH
2 metres (6.5ft)

DIET
Herbivore

PARENT AND CHILD

In 2004, palaeontologists in Liaoning (China) uncovered a group of fossils of Psittacosaurus young with a parent. This was a significant discovery because few fossils of dinosaurs looking after their young have been found.

PARROT HEAD

Psittacosaurus' head and beak were similarly shaped to modern parrots'. However, unlike a parrot, Psittacosaurus also had teeth. These teeth seemed to be towards the back of the jaw, rather than at the front of the beak.

WHIPPING TAIL
PLATEOSAURUS

Out of the thickets comes a young Plateosaurus, searching around for twigs and needles on the forest floor. Suddenly it's attacked by a tiger – a tempting meal to the modern carnivore. The Plateosaurus isn't as slow as its bulk might suggest; it whips its tail around and knocks the tiger off its feet.

Plateosaurus was well-used to the sparse vegetation of the Triassic landscape. In Plateosaurus' own time, the carnivores were small but active **theropods**, like Coelophysis. As well as its whip-like tail, Plateosaurus had a large thumb claw on each of its front hands. It may have been able to fight off primitive meat-eaters, but would it be well-equipped to defend itself against the main meat-eaters today? Possibly not.

PLATEOSAURUS
PRONOUNCED
plat-ee-oh-saw-rus

LIVED
Late Triassic period
214-204 million years ago

LENGTH
8 metres (26ft)

DIET
Herbivore

THE FIRST

Plateosaurus was one of the first big plant-eating dinosaurs to walk the Earth. It was also one of the first dinosaurs to ever be discovered. It was discovered in 1834 in Germany when fossils of several **vertebrae** and leg bones were uncovered.

DINO DETECTIVE

Palaeontologists have debated whether Plateosaurus spent more time on two or four legs. Recent fossil evidence confirms that it could actually only walk on its hind legs. If it still lived today, maybe it would learn to walk on all fours out of survival instinct. Maybe that would give it better balance, and therefore a better chance of fighting off our modern-day predators, like tigers, which are much larger than Plateosaurus' primitive predators.

TERRITORIAL THREAT
ANUROGNATHUS

At a bird feeder, a little bird's meal of seeds is rudely interrupted. A pair of Anurognathus swoop down and chase it away. Although Anurognathus is an insect-feeder not a seed-eater, it's very protective of its territory.

When the dinosaurs were alive, the air was full of **pterosaurs** – flying reptiles. If they were around today, they would have to compete with birds for their food. Like the modern birds, there were many shapes and sizes of pterosaur, each type having a different lifestyle and eating a different food. Anurognathus was the pterosaur equivalent of the modern-day swift – a fast aerial hunter of insects.

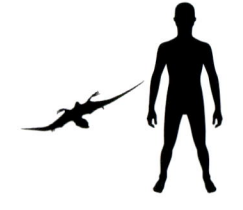

ANUROGNATHUS
PRONOUNCED
an-you-rog-nath-us

LIVED
Late Jurassic period
150-145 million years ago.

WINGSPAN
35-50cm (14-20in)

DIET
Insectivore

WING PRESERVATION
Only about two specimens of Anurognathus have ever been found. However, palaeontologists have a clear understanding of what its wing **membrane** was like because so many of its relatives such as Pterodactylus have been found in the fine-grained **lithographic** limestone of southern Germany. The preservation was so good that the imprint of the skin of the wings was preserved.

BEAK OR JAW?
Different pterosaurs had differently-shaped jaws depending on their diet and lifestyle – just as different types of birds today have differently-shaped beaks.

CHARGE!
STYRACOSAURUS

A white rhinoceros feels the ground shuddering beneath its feet and senses danger. The rhinoceros' instincts are right - with its long horns and heavy muscular body, the adult Styracosaurus that has blundered into its path is a formidable opponent. But the rhinoceros is a fierce rival – it's of a similar size and build, and is desperate to protect its territory. It starts to charge. Several tonnes of muscle are about to collide.

Back in Late Cretaceous times, herds of Styracosaurus likely shared their **habitat** with herds of other horned dinosaurs. They would have been able to identify one another by the horns and other head ornamentation, so the different species of **ceratopsians** could keep to their own territories. The different horn arrangement of the white rhino would certainly signal to Styracosaurus that it's a different animal – to either be avoided or challenged.

STYRACOSAURUS

PRONOUNCED
sty-rack-oh-saw-rus

LIVED
Late Cretaceous period
76-70 million years ago.

LENGTH
5.5 metres (18ft)

DIET
Herbivore

CLEVER HEADWEAR

Styracosaurus' enormous neck frills and multiple horns and spikes made it look much bigger and scarier than it actually was. Their purpose was to scare predators away. However, they also made great weapons if Styracosaurus needed to protect itself or defend its territory if a predator did advance.

BATTERING RAM
STEGOCERAS

A loud crack echoes around a snowy valley in the Rocky Mountains. It sounds like a gunshot but it's actually the noise of two skulls crashing together. A Bighorn ram and a Stegoceras are clashing. It's mating season for the rams - they resent intruders, and fight off any animal they see as competition. Sit tight; this butting contest could last for hours!

Stegoceras would probably be at home in the mountainous areas of today's world. It was well-equipped to deal with hardy mountain animals. Stegoceras was built like a battering ram. Solid bones at the top of its head could absorb the impact of it colliding with a rival – preventing it from damaging itself. A strong, stiff neck and back stopped the animal's body from twisting when fighting, making it an incredibly strong dinosaur.

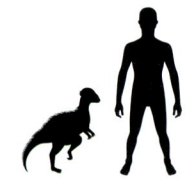

STEGOCERAS
PRONOUNCED
steg-oh-sair-us

LIVED
Late Cretaceous period
76-65 million years ago

LENGTH
up to 2.5 metres (8ft)

DIET
Herbivore

BONE HEAD

Stegoceras was a member of the **pachycephalosaurids** – the 'thick-headed lizards'. They are so-called because of the massive dome of bone on the top of their skull. This could have been used as a battering ram; for head-butting their rivals when competing for leadership of the herd, or for defending the herd from predators. They may have butted one another on the flanks, as well as head-to-head.

SENSE OF SMELL

All pachycephalosaurs had thick, domed skulls which they could use to protect themselves. However, scientists think that they also had a good sense of smell, alerting them to predators in time to run away.

PREHISTORIC PROTECTION

Dinosaurs were protective of their territory, babies and food which meant they often fought each other! Sometimes this meant fighting to the death, especially when one dino was hunting the other as a food source – no dino wanted to become another's meal!

Baryonyx

ATTACKING TEETH AND CLAWS

Dinosaurs would fight with anything they could. Fossils have revealed that many species had sharp teeth and long claws, which not only helped when killing and eating prey, but when fighting for their life, too. The biggest teeth and claws belonged to carnivores, like Deinonychus and Baryonyx, who could pin down opponents with one claw and slash with another. Once in that ferocious death grip, it was impossible to escape.

TOUGH TAIL AND SPIKES

Defensive weapons like bony spikes, horns and strong tails mostly belonged to plant-eating dinosaurs who needed to defend themselves against meat-eating predators. One of the best-defended dinos was Ankylosaurus whose tough spikes and plated body was almost impossible for predators to bite through. Swinging their club tail could do lots of damage, too!

As well as having in-built protection, dinosaurs defended themselves through clever actions, too.

SUPER SPEED

Not all dinosaurs were great fighters; for some, their best option was to run away or hide if threatened. It's thought that some species, particularly smaller plant-eaters like Troodon, were very fast runners. Their small size and super speed meant they could outrun the larger carnivorous dinosaurs who were trying to eat them!

POWER IN NUMBERS

Like modern mammals, many dinosaurs lived and hunted in herds. They were more likely to fight off predators when working together rather than fighting alone. A huge fossil discovery in Patagonia showed the complexity of dino herds. There was a large group of eggs in one area, young dinosaurs in a second area, and adult dinosaurs in pairs dotted around. It seems they were much more collaborative creatures than we originally thought!

THE LIVES OF DINOSAURS TODAY

Both humans and dinosaurs would lead completely different lives if we existed alongside each other. Imagine walking past a field and seeing a group of dinos being herded by a farmer, or pulling a cart full of fruit and vegetables! What do you think would be the biggest change to your day-to-day life?

FIGHTING FOR FOOD

Sharing the world with dinosaurs would be pretty difficult. The biggest dinosaurs had to eat lots of food every day to stay alive. If they still roamed in the wild, they'd likely trample farmland, destroying crops being grown for our food. Wild plant-eaters like Plateosaurus would be huge pests, eating vegetable patches and orchards of fruit trees, whereas wild meat-eaters would target the animals being bred for human food. That's a lot of tough competition to deal with!

DINO DINNER

Smaller plant-eating dinosaurs could be farmed as our main source of food. Traditional livestock such as chicken and sheep would be classed as wild animals and rarely seen, like deer today. Instead, you would see herds of dinosaurs, such as Psittacosaurus, roaming the fields – and even see them on your plate!

NEW HOMES

There would be much less space on Earth if we shared the world with dinosaurs. Our city skyscrapers would have to be built much taller to take up less land space. Dinosaurs would come up with new ways of living, too. We might see the smaller species learn to burrow and build homes underground, as modern lizards, snakes and foxes do.

DIFFERENT SPECIES

Who knows how many dinosaur species would have existed. We may have entirely new dinos roaming, as well as variations of existing species. We know that Triceratops and Styracosaurus had different arrangements of horns on their heads – just like the modern rhinoceros. Would they eventually have grown different numbers of horns, or other new features to adapt to living in different areas?

DID YOU KNOW?

Dinosaurs are fascinating creatures. Scientists are constantly discovering more about them and finding answers to the world's most curious questions. Did you know these amazing facts about dinosaurs?

HOW DID DINOSAURS FIGHT?
With teeth, horns, spikes, tails, claws – with whatever they had!

IF DINOSAURS SUDDENLY APPEARED TODAY, HOW WOULD WE LOOK AFTER THEM?
In a very big zoo! Most dinosaurs were very large. The bigger an animal is, the bigger the area it needs to live in. We would also need vast farms to produce enough food to keep captive dinosaurs alive.

WERE THERE ANY DISEASES THAT COULD KILL DINOSAURS?
There must have been. Palaeontologists have found deformities in bones that were obviously caused by diseases which resulted in infections, cancerous tumours and arthritis!

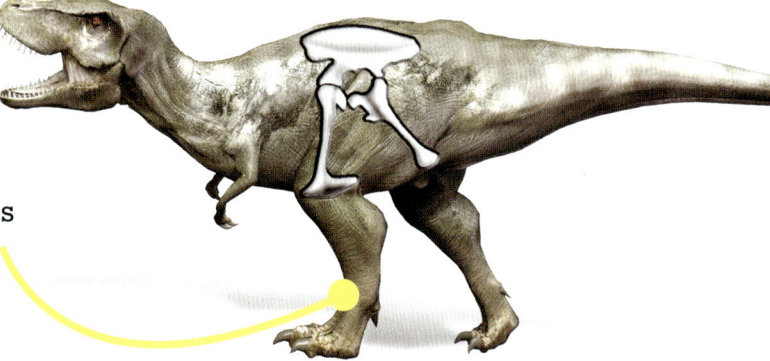

WERE PREHISTORIC PARASITES MORE DEADLY THAN MODERN PARASITES?

The thing about parasites is that they only attack a specific host. So parasites in dinosaur times would have been as deadly to them as ours are to us.

WHAT WERE THE FIRST AND LAST DINOSAURS TO LIVE ON EARTH?

For years, Eoraptor held the title of oldest dino. But it's recently been beaten by Nyasasaurus, who lived 243 million years ago! At the end of the Cretaceous period, all the dinosaurs that still existed died out together – so it's hard to say what was the last. We do know that they included Tyrannosaurus, Ankylosaurus, and Triceratops.

HOW MANY SPECIES OF DINOSAURS WERE THERE?

We have discovered over 700 species, but more are being found all the time.

TRUE OR FALSE?

There are a lot of fun facts about dinosaurs but what is actually true and what is a myth? Put your dino knowledge to the test with this true or false quiz.

DINOSAURS LAID EGGS!

TRUE. All the dinosaurs discovered so far reproduced by laying eggs rather than giving birth to live young. From fossil evidence, it seems likely that different species laid different numbers of eggs at a time.

DINOSAURS MOVED REALLY FAST!

FALSE. Although some dinosaurs, like Struthiomimus, were fast movers, some of the biggest dinosaurs were so huge and heavy that they would have been much slower. Scientists think that T.rex moved at the speed of a walking human.

RHINOS ARE DESCENDED FROM TRICERATOPS!

FALSE. Even though they look like one another with their big bodies and horns, they're very different animals. The reason they look similar is because they have a similar lifestyles and must defend themselves against predators, hence the large horns. However, Triceratops was classed as a reptile, whereas rhinos are mammals and a lot smaller than the dino in question!

WE DRINK THE SAME WATER DINOSAURS DID!

TRUE. The water on Earth has been here for millions of years. Thanks to the water cycle which sees the same water evaporate and rise into the **atmosphere**, cool to form clouds, and then fall as rain over and over; we're constantly drinking and recycling water the dinos drank!

THE LONGEST DINOSAUR NAME IS MORE THAN 20 LETTERS LONG!

TRUE. It's Micropachycephalosaurus, which has 23 letters!

ALL DINOSAURS WERE COVERED IN SCALES!

FALSE. For a long time, palaeontologists thought all dinos were covered in scales like modern-day lizards. It wasn't until the 1990s that they discovered evidence that some dinosaurs had feathers. We now also know that others, like Anurognathus, had furry fibres on their wings.

UNCOVERING THE PAST

Can we see live dinosaurs today? Yes, if we count birds as dinosaurs. No, if we're thinking about the big dinosaurs in this book. In this case, we must rely on fossil evidence to tell us what they looked like. Since the 1820s, when the first dinosaur fossils were found, we've been building up our knowledge piece by piece (literally).

Our knowledge of dinosaurs is far from complete: it's very rare for scientists to uncover fossils. Not only are fossils usually deep underground, but it's rare for land-living animals to fossilise in the first place. To fossilise, the animal needs to have been buried quickly, otherwise bones get scattered and rot away from exposure to weather and **bacteria**.

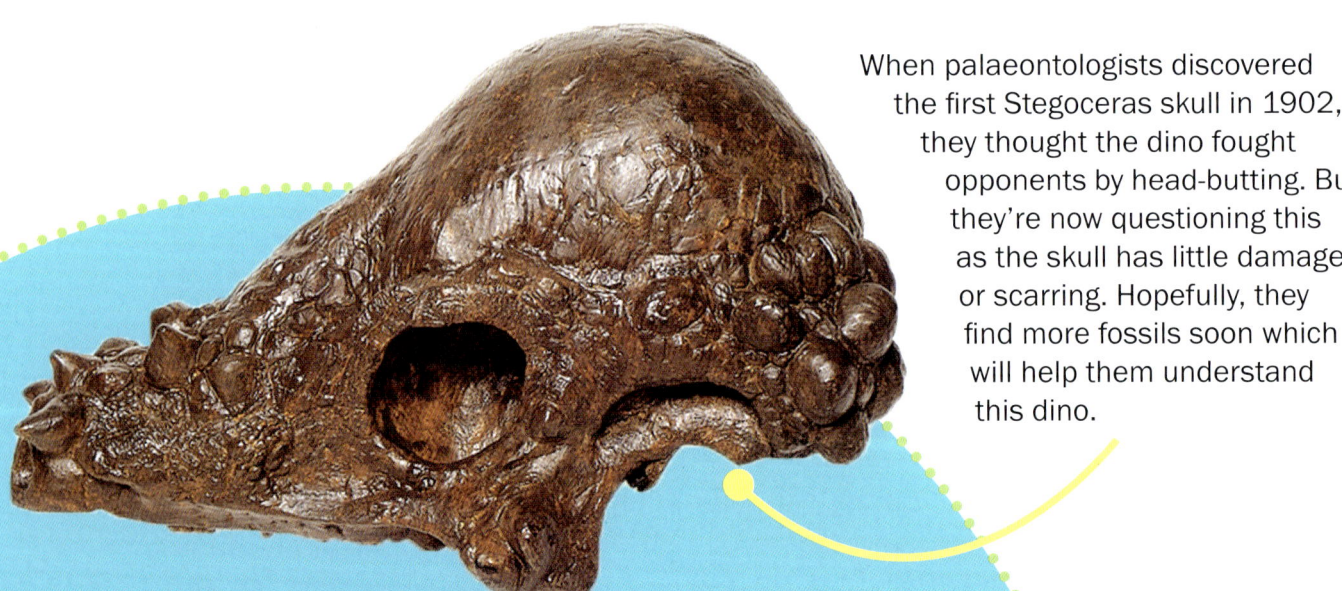

When palaeontologists discovered the first Stegoceras skull in 1902, they thought the dino fought opponents by head-butting. But they're now questioning this as the skull has little damage or scarring. Hopefully, they find more fossils soon which will help them understand this dino.

Mosasaurus tooth

Fossils can be discovered in different circumstances. Usually, we only find isolated fossil bones which have been separated from the rest of the skeleton. They're not entirely useless - even an isolated tooth or limb bone may be identifiable to a species. However, because the bone has spent millions of years in the ground, it's often weathered or fragmented, making it tricky to find out which dinosaur it came from.

In the early 1900s rush to make new discoveries, palaeontologists over-split dinosaurs into lots of species, even when fossils only showed tiny differences. At one point, there were three species of Styracosaurus, but re-examination has found that these fossils belong to one species.

Only very occasionally are articulated skeletons uncovered – this is when the bones are still joined together, as they were in life. More commonly, associated skeletons are found – this is when the bones are jumbled up, but it's obvious that they came from the same animal. It takes a very knowledgeable scientist to put the bones back together.

Articulated skeleton of a velociraptor.

ANYONE CAN FIND A FOSSIL!

You don't need to be a professional palaeontologists to discover a prehistoric creature. Many have been found by children. In 2021, a young girl discovered a 220-million-year-old dinosaur footprint! Given that dinosaurs appeared 230 million years ago, this footprint must be from one of the earliest dinosaurs to walk the Earth.

Perhaps the greatest fossil hunter of all was 12-year-old Many Anning, who found and **excavated** the first complete skeleton of the prehistoric marine reptile **Ichthyosaurus** in England in 1811.

Why not begin your own searches by joining a fossil-hunting group?

Statue of Mary Anning, fossil hunter in Lyme Regis, Dorset (England).

INDEX

A
Ankylosaurus 9, 12-13, 22, 27
Anning, Mary 31
Anurognathus 8, 16-17, 29
asteroid 6

B
Baron Georges Cuvier 10
Baryonyx 22
beak 9, 13, 16
bird(s) 6, 16-17, 30
bone 13, 20-21, 26, 30-31
bone plated 12-13

C
carnivore (meat-eater) 7, 14, 24, 26-27
Cretaceous period 8-9, 10, 12, 18, 20, 27

D
Deinonychus 22
disease(s) 26

E
eggs 23, 28
Eoraptor 27
extinction 6, 9

F
farm 12, 24, 26
fossil 6-7, 10, 15, 16, 22-23, 30-31

H
herbivore (plant-eater) 10, 23, 24
herd 7, 10, 15, 18, 21, 23, 24
horn(s) 7, 9, 18-19, 20, 22, 25, 26, 28

I
insectivore 16

J
Jurassic period 8, 16

M
Micropachycephalosaurus 29
Mosasaurus 31

N
Nyasasaurus 27

P
Pachycephalosaurus 21, 28
Plateosaurus 7, 8, 14-15, 22, 24
parasite(s) 27
Psittacosaurus 9, 10-11, 24

S
sheepdog 12-13
sifaka 10-11
Sir Richard Owen 6
skeleton 15, 31
 articulated skeleton 31
 associated skeleton 31
skull 9, 20-21, 30
Stegoceras 9, 20-21, 30

Struthiomimus 28
Styracosaurus 9, 18-19, 31

T
tail 8-9, 12-13, 14-15, 22, 26
teeth 11, 22, 26, 31
tiger 7, 14-15
Triassic period 8, 14, 27
Triceratops 8, 25, 27, 28
Troodon 23
Tyrannosaurus rex 8, 27, 28

R
ram 20-21
rhino 7, 18-19, 25, 28

Z
zoo 26

GLOSSARY

Atmosphere - The layer of gases which surround the Earth.

Bacteria - Microscopic organisms that can cause disease.

Carnivore - An animal that feeds on meat.

Ceratopsian - A type of dinosaur that had spines, horns, and a bony frill on its head.

Excavate (verb) - The careful removal of earth from an area in order to find buried remains.

Fossil - The remains or impression of a prehistoric plant or animal embedded in rock and preserved.

Habitat – The place or environment where a plant or animal naturally lives or grows.

Herbivore - An animal which eats only plants.

Ichthyosaur - A type of swimming reptile from the Mesozoic Era. They had streamlined fish-like bodies with fins and a tail.

Insectivore - An animal which eats mostly insects.

Juvenile – An animal which is young - not yet a full-grown adult.

Lithographic – Something which can be imprinted onto a plain surface.

Mass extinction - An event that brings about the extinction of a large number of animals and plants. There have been about five mass extinctions in the history of life on Earth.

Membrane - A thin soft layer of material.

Meteorite - A rock from space.

Mesozoic Era – The era of time in which dinosaurs lived, among other animals. It lasted around 186 million years, from 252 to 66 million years ago.

Oases - Green areas in a desert, where there is water and plants grow.

Pachycephalosaurid – A type of dinosaur that had a very thick skull roof.

Palaeontology - The study of ancient life and fossils. People who study this are called palaeontologists.

Period - A division of geological time that can be defined by the types of animals or plants that existed then. Typically, a period lasts for tens of millions of years.

Prosauropod - A group of dinosaurs from the Late Triassic period who were the ancestors of sauropods – the long-necked, plant-eating dinosaurs.

Pterosaur - One of a group of flying reptiles from the Mesozoic Era. They flew with leathery wings supported by an elongated finger.

Theropod - A type of bipedal carnivorous dinosaur with long jaws, three-toed hind legs, and small front legs with clawed hands.

Tsunami – An extremely long and high fast-moving sea wave caused by an earthquake or other disturbance.

Vertebrae – The bones that make up an animal's spine.